101
WATER
WISE
WAYS

101
WATER
WISE
WAYS

HELEN MOFFETT

BOOK**STORM**

ISBN: 978-1-928257-54-7
e-ISBN: 978-1-928257-55-4

First edition, first impression 2018

Published by Bookstorm (Pty) Ltd
PO Box 4532
Northcliff 2115
Johannesburg
South Africa
www.bookstorm.co.za

Edited by Russell Clarke and Paige Nick
Proofread by Wesley Thompson
Cover and book design by mrdesign
Author photograph by Martine Bisagne
Images and graphics by iStock
Printed in the USA
For more information on Helen Moffett, visit
www.helenmoffett.com

For Justin and Fiona,
the Well Angels,
with thanks from the
Water Lily.

WHY OUR RELATIONSHIP WITH WATER NEEDS TO CHANGE

Cape Town is in crisis: after years of drought, it faces the very real possibility that its taps will run dry as dams and reservoirs drop below critical levels. The threat of Day Zero – switching off municipal water supply, after which residents will have to queue for a daily allotment of 25 litres per person – hangs over the city. Is this a once-off? Watergeddon? Will the winter rains come in time to save us?

The reality Capetonians face now is going to keep rearing its head. We're using too much water at the same time that supplies are starting to dwindle for a whole range of reasons (climate change, for a start). We may get lucky this time – we might still get good winter rains – but this issue isn't going away.

But this isn't a reason for doom, gloom and despair. It's a unique opportunity for everyone to rethink the way we interact with this most precious resource: from the way we wash ourselves and our clothes, eat, cook, clean, use the toilet, to how we build our houses, design our cities and grow our food in the future.

Cape Town is famous for its beauty, infamous for its history, plagued by a host of social ills and wounded by extreme inequality. How its citizens act in the face of this crisis could become a model for all cities, around the globe, as they also start to struggle with water scarcity.

We live in an arid country on a planet with finite resources. As I write this, three provinces in South Africa have been declared national disaster zones because of drought. The way we think about water needs to change, and fast. This is especially true for those of us who have running water and flush sanitation piped into our homes. For millions of South Africans, water is already a precious resource that costs toil to collect and fuel to heat. Our middle-class expectations that water will gush steaming from our dozens of indoor taps 24/7 are going to look as bizarre to future generations as the spectacle of Cleopatra bathing in asses' milk. Our Roman-orgy relationship with water is over.

There's no doubt about it, our grandchildren are going to think we're insane:

Wait, you kept huge bodies of water in your backyards just so you could swim in them for a few months of the year? You peed in drinking water? You showered or bathed every day? You had bedrooms with bathrooms attached? What do you mean, architects were allowed to design houses where gutters just carried water away into the ground? You had garages for your cars, but no water tanks? Were you people completely crazy?

One problem is that for good reason – everyone needs it for life, health and dignity – water is incredibly cheap. Access to water is universally recognised as a human right. Without it, we die. What's more, even if we have water, unless it's clean and safe to drink and use, we also die, as many cholera epidemics and the movie *Erin Brockovich* taught us. This is why, with a few exceptions, providing water is the job of governments and municipalities, which are required to keep the cost as low as possible. However, the fact that our treated water costs pennies, in a world that respects cold cash, means that we don't value it.

Another issue, and this is true for many countries, is that we have a system in which water treated so that it's safe for drinking and cooking is used for everything: we toss this life-giving and life-saving resource casually onto our gardens, into our pools, down our toilets. Think of the poster of the small African child asking the Western aid worker, "You have so much clean drinking water, you POOP in it?"

This leads to the next problem: water is linked to some of our most private bodily functions – defecating and urinating, menstruating, even vomiting. In the last two generations, in most Westernised cultures, we've been able to take these unavoidable biological processes behind closed doors and to our porcelain thrones. Water makes everything "nasty" and "messy" go away. We flush, and forget about it. Thinking about the water crisis means we have to face these earthy realities, and we'd much rather not.

In this book and in our water-wise future, we're going to have to be tough in facing these realities. We're going to have to consider what to do when we have a tummy bug, or a period, or an incontinent family member. Apologies in advance if this causes offence, but I will be using the word "shit" to describe, well, shit – this is no time to be dainty about No 2s, "poo" makes me think of nursery school, and "faeces" is a bit intimidating, as well as tricky to spell.

The recent panic buying and illegal stockpiling of water is likely related to real terror at the prospect that come Day Zero, Capetonians will not be able to flush their toilets. Many have realised, with dismay, for the first time, just how much water this uses: 5 to 9 litres every time we touch that handle.

This book will hopefully help to alleviate water panic and distress. A "can-do" compendium, it's meant to be a guide, not prescriptive – not all solutions or tips are one-size-fits-all. Think of it as an ally in your fight to save water, and part of your survival kit, along with the first-aid box; Valium for water-worriers.

In it are the best tips found so far: from the tiny (but every bit helps) to the huge and long-term. Some tips will cost money to implement; many are free. Some depend on individual circumstances (whether you have a garden, for instance); some apply to everyone. You'll find hints on everything water-related, from ingenious new ways of showering to laundry short-cuts, from practising good water hygiene to building your own dry urinal. There are tips for kids, families, elders, guests, tourists; for homes, kitchens, offices, businesses, institutions. And these are just a start: refine them, adapt them to your specific situations, and add to them.

This is a marathon, not a sprint; we may even find that we're in for the Comrades. So don't be surprised if you get water fatigue. Don't worry if you struggle at first to adjust; your new water-wise ways will soon become second nature.

Be aware that the way the middle classes are going to have to adapt is already daily life for millions of South Africans. Many queue for scarce water at standpipes and wells, and bear the burden of carrying it home, storing it and heating it. For some of us, it's a humbling experience to consider that Day Almost Zero is reality for countless families.

We're at a crossroads. This is an unprecedented opportunity to become better neighbours, stronger communities and more close-knit families. There's huge satisfaction and pleasure to be gained from coming together to solve a communal problem. Remember, we will get through this. Mostly by being good neighbours (much more on this later – see pp. 73–79). We are not about to descend into anarchy and chaos.

This crisis also offers opportunities for new entrepreneurs, inventors, innovators and businesses: everywhere on the internet, we see great new MacGyver-type contraptions for saving or collecting water, from water-harvesting umbrellas to wheeled water-trolleys. Businesses that collect, process and recycle "humanure" (the contents of composting toilets) are going to boom in the near future – these are the start-ups to be buying shares in.

The lessons we can teach our children in this moment are huge: if they see us "taking care of business", getting out tools and adapting our households and appliances instead of wringing our hands or blaming others, they'll feel more optimistic and less helpless. Even more important, if they see us modelling behaviour in which we look out for the weak and the vulnerable, show consideration for others, put the common good before our own desires, and look for practical and even fun solutions to problems, we'll be raising future world and South African citizens who will be wise, water-resilient and compassionate.

Now turn the page and dive in to the first of 101 (and more) water-wise ways. Welcome to the future!

Helen Moffett
Cape Town, March 2018

PS:

We'd love to hear your best water-wise tips. Tweet them to @BookstormZA.

GETTING YOUR HEAD INTO GEAR

TIP 1

Don't panic! We're not the first city in the world to run out of water – the huge Brazilian megacity of São Paulo endured grim months of little or no water between 2015 and 2016. We're not even the first South African town to do so: that privilege belongs to Phuthaditjhaba, which ran dry in October 2015 and had to rely on water tankers for months. And Johannesburg had a near-miss two years ago, saved by good rains and the Lesotho Highlands Water Project. But Cape Town is the most visible – the first major world tourist destination to do so. The eyes of the world are watching to see how we'll cope. So take a deep breath, and keep on carrying on.

TIP 2

Go Green. This is probably the most important short- and long-term strategy for becoming water-wise, and this is a subject for a dozen separate books. Many water-saving tips belong within a set of broader green-living principles, along the "reduce, re-use, recycle" lines. We do things (like throwing away food) that waste the water used for growing/production in the first place – but we're not taught to think of the secondary consequences of such everyday acts.

So a general water-saving rule of thumb is to follow an environmentally friendly lifestyle. (This also has the obvious benefit of leaving a habitable planet behind for your children.) If you're new to this, make changes little by little – you do not have to rush out to buy hemp sandals and an electric car right this minute. Generally, though, the greener your lifestyle, the less water you'll use.

Green living **saves** money. It does not mean buying every expensive product in the health shop – it means re-using goods and using less of them. You'll be delighted at the effect on your pocket.

TIP 3

However, the urgency of this crisis also means we'll have to temporarily sacrifice some green principles – those of us with cars will be using them a lot more to haul water, we're going to be using more disposable plates, napkins, nappies, and so on. So for the duration, accept that you may have to compromise on some of your principles. (The person who plans on using only glass bottles when collecting water because she's opposed to plastic is looking at either a ruptured disc or a riot.)

TIP 4

Ditch the magical thinking. We're all holding out for "good rains" – as if that will miraculously make this whole problem go away. IF the Western Cape gets good winter rains, and IF they fall in the right places, and IF they're the right kind (we need soft and soaking rain rather than sudden cloudbursts that rush away down drains and out to sea), we will indeed be off the hook – temporarily. But a major underlying contributing factor, climate change – extreme weather conditions caused by global warming – is not going away anytime soon. We're likely to face this scenario again and again in the near future.

Lessons learned from the São Paulo drought of 2015: residents cut their usage dramatically, largely in response to financial incentives (something for our municipalities to consider), but after good rains, at least half their households went back to their old ways, and the city is once again teetering on the brink of water collapse. Changes need to be permanent. This is something for Jozi dwellers to consider, too.

Make water-saving a family project, as opposed to one more thing you nag the kids about. Have a weekly water indaba, and encourage everyone to come up with their own tips, and to tweak family practices. Share, encourage and praise water-wise habits. Set your goals for water usage (now 50 litres a day, which will drop to 25 litres if we hit Day Zero), and use the water family meeting to do the sums and track how you're doing. Send out for pizza if goals are met, or give a small prize to the family member saving the most or with the best innovation.

If you spot a burst or leaking pipe, contact the authorities immediately. Find out what the numbers are for your location and enter these into your phone. The numbers for Cape Town are on the right.

Call:
0860 103 089
SMS: 31373
WhatsApp:
063 407 3699

Then put on your gumboots, fetch your containers and scoop up that water, if it's safe to do so.

This is a learning curve. Water-saving techniques we were using a few months ago now seem wasteful (remember when we poured grey water onto plants instead of saving it for flushing?). Keep adapting and refining processes, and encourage the whole family to do so.

The water crisis can feel like an endurance test at times. Remember, it won't always be this tough. Our water-wasting days are over for good, but the current need for water austerity will relax in time. Think of it as a diet in which we have to permanently change our eating habits: there's a very strict initial stage, then a stage where we expand the menu a bit, and finally a maintenance stage in which we can eat the occasional slice of cheesecake. There will be baths in our future, but as a special indulgence, not a daily routine.

Shirley Conran wrote that when undergoing a sudden huge change in standard of living (bankruptcy, eviction, running out of water – OK, I added that last one), try to hang onto one small regular treat or luxury. Having something to look forward to boosts morale tremendously. See if there's a responsible way to experience a little water bonus now and again: mine is visiting the hairdresser, taking a few bottles of clean but undrinkable water from a local spring to cover washing and rinsing.

Setting up water-harvesting systems is one of the most effective ways to feel more secure in the face of the water crisis. This is especially true if you can afford to do so; spending a bit of capital on water infrastructure by adapting your guttering, installing tanks or a well, pumps and a grey-water recycling system will give you independence, the ability to help those with fewer resources, and improve the value of your property.

TIP 10

Water harvesting means thinking about water storage. It's all very well collecting hundreds of litres in buckets and bins when it rains, but where are we going to put it? Likewise, if after Day Zero, tankers deliver to NGOs and vulnerable households as planned, where is this water going to be kept, and in a hygienic state?

Rainwater tanks are an obvious solution, but they're not cheap, require space, and in the Cape Town area, there are long waiting lists to buy them. Get creative: use the bath, dustbins, Portapools, special water "bladders", flexipipes (roll up the open ends and seal with pegs) and more. "Hippo" rollers can be moved around easily, even by children and older folk, and take up little room. They are not cheap, but are durable, and can be donated to an NGO if you no longer need them.

NB! Storing water does NOT mean getting huge tanks and then filling them with municipal water (which will be unusable for drinking purposes within a week), or buying 100 litres at a time from a water shop (which means two days' municipal water allowance gone before you've even had a cup of tea). Use your tanks for rain-harvesting by connecting them to the gutters and downpipes of your home. (Some institutions and buildings have no choice but to install tanks for storage purposes; I'm talking about private dwellings.)

TIP 11 **If you have limited space, but cash to spare, get rain chains** attached to your gutters and run them into those elegant tanks designed to look like planter pots. This would be a good option for those with only balconies or small courtyards or patios.

TIP 12 **Take care of your body and get strong.** Heaving buckets takes muscles. Talk to a physiotherapist, yoga teacher, athlete, dancer or Pilates instructor about how to develop core strength and lift your full containers without giving yourself a hernia – doctors are seeing a rash of folk hobbling into their offices with "bucket back". Use only containers you can lift once they're full.

TIP 13 **If you know you're going to be vulnerable in case of Day Zero –** your arthritis means you can't hoist buckets, you're a single parent with small children, you have no private transport – reach out now to your neighbours and friends. Think of how you might be able to offer services or favours in exchange for help with water. You could cook meals or walk the dog for those willing to collect water for you; or offer babysitting or pet-sitting services. Set up playdate rosters in your homes so that other parents can take it in turns to fetch or harvest water.

GETTING
TO GRIPS
WITH WATER

Knowledge is power. Learning about and taking charge of our own water needs instead of sitting back and hoping the whole thing will go away will make us feel stronger and more confident. Get informed about the water around you, how to use alternative sources safely, how to store it, how to recycle used water, and just how little you can get by on. This process of getting "water fit" will make you feel prepared for almost anything. Except perhaps the zombie apocalypse, and we're working on that.

Find out if there is a spring nearby, and what the status of that water is. Put pressure on your local councillor to provide reports on the chemical and bacterial profile of the water, and to regularise access (in terms of parking, security, fair use, etc). Collecting spring water will almost certainly require that you have your own or shared transport, although if there is one very close to where you live, adapt a wheeled suitcase to hold your containers and set off on foot. Your collection system will depend on the layout of each spring: some folk are literally climbing mountains or getting down into ditches to collect this water. Wear appropriate shoes and clothing.

Spring collection etiquette tips:

→ Stick to allotted limits (the unofficial amount seems to be 25 litres). If a lot of people are using the spring and you need more than that, return to the back of the queue.

→ Take appropriate containers and don't cause bottlenecks. I once watched a couple hold up a line for half an hour as they collected a vast amount of water in 2-litre cooldrink bottles.

→ Be considerate about parking and don't block roads or driveways.

→ Be relentlessly polite and friendly. Introduce yourself and ask people their names. Help if you spot any opportunities: the optimum number for harvesting is two, one to hold the containers under the spigot, the other to twist lids on and off and pass containers. If you help others, it seems to create a ripple effect, with strangers teaming up for faster harvesting.

→ Throughout the world, women walking to fetch water have sometimes been targeted for sexual attack: a horrible reality, but one to be aware of when collecting from springs on foot. Travel in groups if possible.

BE
RELENTLESSLY
POLITE
AND FRIENDLY

**Get to know your water –
what can and can't be used
for drinking:**

TIP
15

→ **Bottled water:** 100% safe to drink. This includes sealed water from shops selling filtered water. If you're refilling used containers at these outlets, you shouldn't stockpile this water for drinking – see Tip 16.

→ **Municipal water:** 100% safe to drink, although you might like to filter it to improve the taste.

→ **Spring, well and borehole water:** these may SOMETIMES be safe to drink. Get as much information as you can about the source of your water. There are two main kinds of contamination to worry about:

→ chemical pollution and heavy metals; in an emergency, you could run this water through a special filter – see Tip 17, but generally, do not drink this water ever;

→ bacterial contamination; the good news (although not for your electricity bill) is that this is safe to drink after boiling for three minutes, or you can use water-purification tablets. Note that tablets and boiling will solve bacterial contamination; they cannot cleanse water that has heavy metals in it or has been polluted by chemicals. Save the latter for flushing.

→ **Untreated/unfiltered rainwater:** depending on the state of your gutters or harvesting system, this can range from visibly dirty to clean and even drinkable once boiled. Note, however, that roofs, while efficient surfaces for gathering water, are often made of materials that aren't meant to be ingested.

→ **Grey water:** any kind of water that has already been used for laundry, personal washing or household cleaning, and has detergents and other substances (soap, bleach, etc) in it – it cannot be drunk, and should be reserved exclusively for flushing.

→ **Black water:** any water that has (potentially infectious) biological material in it. This includes water in which there are human bodily wastes or fluids, or food – for instance, in which you've washed a cut or cleaned a menstrual sponge or mopped up after a vomiting child. Use this only for flushing.

    → Black water also includes almost all washing-up water, especially if the latter has grease in it. If this water is greasy (has fatty globules floating on the surface), **do not use it for flushing**; this should either go in a special "trap" in the garden (see Tip 27) or be poured down the kitchen or scullery sink. It is much easier and cheaper to clear a clogged kitchen-sink U-bend than a blocked toilet.

TIP 16

Practise good water hygiene – your health depends on it. Any water we collect or harvest needs to be stored appropriately and used in such a way that it poses no hygiene risks. Water is life: it's also a frighteningly efficient means of incubating and spreading disease. Water shortages lead easily to outbreaks of gastro bugs, some of which can be lethal. Coping with vomiting and diarrhoea without running water is a nightmare scenario.

→ Water that looks clean but which has or may have bacterial or chemical taint is excellent for washing yourself, your clothes, your dishes, and general cleaning around the home.

→ If you have small children, elders, invalids or immune-compromised people in your home, use only bottled and municipal water for drinking and cooking.

→ Some borehole, well and spring water is perfectly safe to drink, and often delicious: the Newlands springs in Cape Town are popular for this reason. However, be aware that your storage systems need to be impeccably hygienic; your containers need to be sterile, and the harvesting process (which can involve mud, dust and bits of vegetation flying around) needs to be as squeaky-clean as possible.

→ Practise strict segregation between drinking water and non-potable water (put masking tape labels on your containers or write on the sides); also keep grey water far away from the rest, and black water the furthest away of all. The latter two pose the most immediate health risks. Use rubber gloves when handling grey water that's been sitting around for a while.

→ Segregate your water containers; reserve the ones that can withstand boiling water for drinking water only, and keep them spotless. As soon as they're empty, add a spoonful of bicarbonate of soda or Milton, a cup of boiling water, shake, empty, add a bit more boiling water and rinse. I also pour boiling water over the spout and lids. Keep and re-use those 5-litre bottled water containers for storing non-drinkable water. They're more convenient for refilling toilet cisterns than open buckets.

→ Assemble your scrub buddies. Bicarb is everyone's new best friend, as is spirit vinegar. These safe and non-toxic "natural" disinfectants are good for use in water containers. Along with these gentle aids, good old deadly bleach is handy; I use the thick kind for the toilet bowl, and the thin kind to sterilise kitchen and bathroom cloths.

→ The easiest and least messy way to fill a toilet tank is with a watering can.

→ Kitchen cloths can become bacteria mosh pits in current conditions, so at regular intervals, pour a little boiling water on them and add a splash of bleach. Sterilising fluids like Milton are also useful.

TIP 17 — **You'll find yourself continually boiling water,** which is not the greenest thing to do, especially if your kettle is electric. Those who cook with gas will already know that heating water this way is cheaper and better for the planet. It might be worth investing in a gas camping stove. (See Tip 44 for more water-heating options.)

Bonus tip: you can buy water filters with ceramic "candles" that render water with heavy metal and most chemical pollution safe to drink. These are heavy and expensive (they are not the same as jug or sink filters installed to make tap water taste better), but worth the investment if you have borehole or well water that has heavy metals in it, but is otherwise clean.

TIP 18

It's vital that you keep your hands clean, and ironically, you might find yourself washing them far more often than usual. Keep a bowl of warm water with disinfectant soap in the kitchen when cooking and cleaning and use it frequently; another option is a small spray bottle with a little boiled water and a dash of Milton in it: good for squirting onto hands and much cheaper than hand sanitiser, which can be hard on the skin. (See Resources for how to make your own wet wipes.)

Bonus tip: another good way to keep your hands clean is via the bottle-plus-straw method. Take any plastic bottle with a lid. Punch a hole close to the bottom with a knitting needle and insert a plastic straw. (Some seal the join with a ring of Prestik.) Fill the bottle with water and stand it at the sink or basin so that the straw juts out. To wash your hands, or a piece of fruit or veg, press the bottle and water will trickle out the straw. If you need a stronger flow to get everything clean, twist open the lid. This works like a charm.

TIP 19

You can wash veggies and fruit in clean non-drinking water, but add a sprinkle of Milton. The latter and similar sterilising fluids are also good for wiping down kitchen surfaces where germs must be headed off at the pass (fridge shelves, chopping boards, microwave interiors, etc). Remember: tummy upset + water shortage = perfect storm of misery.

TIP 20

Practise "water-fasting": seeing just how little you can manage on for several days. This will make you feel far more in control, and will help you develop a "reserve" for times when you need extra water: for example, when experiencing a heavy period, doing particularly hard or dirty labour, struck by conditions like diarrhoea, night sweats, eczema, and so on. Most healthy adults find they can live comfortably on as little as 30 litres a day (including laundry, cleaning and water for animals), and much less for short bursts. If you can, use less than your allotted 50 litres a day, so that those who need the extra – small children, invalids, the incontinent and so on – can benefit. Within families, parents can use less daily water so that their children or older relatives can have a bit more.

Experimenting, I found I could get my water consumption down to 10 litres a day: 3 litres for drinking and cooking (including drinking water for my pets); 2 litres in a bucket for a sponge-bath and for handwashing the day's undies (all saved for flushing); 2 litres for doing dishes, washing my hands and keeping the kitchen clean (if you can keep grease out of this water by licking plates and scraping pans, it can also be used for flushing); an extra 3 litres for flushing (I used dirty rainwater, taking daily consumption down to 7 litres, but not everyone has this option). This was surprisingly easy, but keeping it up for a long time would be the real challenge.

TIP 21

Stock up on a small amount of bottled water for emergency drinking and cooking use. At the same time (ironically), we should stop drinking it if we have access to safe tap water. This is because it takes far more water to "manufacture" bottled water than just the contents of the bottle. Plus a lot of it is basically expensive tap water dressed up in planet-choking plastic. If you don't like the taste of tap water, filter it, and if you like bubbly water, invest in a Sodastream dispenser.

Bonus etiquette tip: take your own bottled water to parties. Keep your eyes averted from your host/ess's supply and your covetous thoughts to yourself.

TIP 22 **Stock up on medical supplies needed** for stomach bugs or bladder infections – anything that might have you or your family members trotting more urgently to the toilet than usual (anti-emetic, nausea and diarrhoea meds, rehydration salts, Citro-Soda, cranberry juice, probiotics, activated charcoal tablets, and so on). Perhaps pharmacists could make up and sell #WaterCrisis packs of over-the-counter medicines for tummies and bladders. It would be good to have the lot in a single basket.

TIP 23 **Add water-purification tablets and sterilising fluid to your emergency pack,** not because frogs are about to start spawning in the city's scant remaining supplies, but because it can be hard to keep supply lines spotless when handling containers, hoses and buckets – especially when the wind is kicking up dust.

TIP 24 **Join a water-saving or water-shedding forum** if you're at ease with social media. Note that while these are excellent sources of information, tips and ideas, you will likely have to wade through a lot of grumbling, blame-mongering, repetition and wilful ignorance. Worse, you may encounter conspiracy theories, fake news, political harangues, reprehensible outcrops of racism and other variants of discrimination. If you are unsure of information posted on these sites, especially if it seems alarmist, double-check the topic via Google and Wikipedia.

Bonus tip: this is an excellent opportunity to teach yourself research skills. Learn how to distinguish between reliable information and sources and those that are dodgy. If a scientific claim is made by several reputable institutions, but denied by one survey of 17 people published by the Snake-Handlers' College of Toadswamp, Alabama, I wouldn't take any bets on the latter being correct.

A major bone of contention on water-saving social media forums is the role of religion. Let's respect all holders of all faiths, including those who believe that prayer will bring rain. At the same time, one of the most sincerely devout people I know says: consider it possible that God intends for you to be the answer to your own prayers – and then go out and get busy loving your neighbour.

WATER-PROOFING YOUR HOME & GARDEN

If you have a garden, consider yourself lucky. This is going to be a great ally in your water-wise mission. Any kind of outdoor space will help, especially if it has a washing line and place to store containers.

Dig a compost pit. It may sound off-track, but this will save you water. There are many composting systems, some involving earthworms, special containers (these are good for tiny gardens with little accessible soil), rotating drums and more. I simply dig a hole about a metre deep and a metre across, and dump everything biodegradable in it. Why? There are a thousand excellent reasons to keep garbage out of landfills and to feed organic matter back into the soil, but for now: it will save washing up. If you are shaken by the notion of licking your plate, or getting the family dog to do so (see Tip 64), then scraping your plate into the kitchen compost bucket after meals is the next best thing.

I compost fish bones and skin – tomato plants love these – and also chicken bones, but I can get away with this because it's a rare occurrence. Generally, meat bones should be kept out of compost heaps unless you want visits from neighbouring dogs; bread might likewise attract rats. Consider thrifty ways to use leftovers – chicken carcasses for stock, and so on (see Tip 61) – or resort to the dustbin.

A compost pit is also a suitably earthy place to dispose of blood (from a mooncup, for instance, or biodegradable sanitary pads) or vomit. Sprinkle a good layer of soil or mulch over afterwards. Note that urine is good for compost heaps, but for reasons too complicated to go into here, this is not a safe place to dump your dump.

Dig a small, deep fire pit in which to burn certain kinds of refuse: food-soiled paper and cardboard (napkins, pizza and cake boxes), "pee" paper, used wet wipes and so on. Don't burn any form of plastic or polystyrene. Obviously, proceed with extreme caution when lighting any fires: you don't want to burn down your house or the neighbourhood.

Construct a home-made grey/black water filter if you have a veggie garden or plants you want to keep alive: Google will give instructions, but I made a small brick enclosure in my garden and layered stones, broken bricks and chunks of wood into it, then topped it with gravel, sand and mulch. This can receive your black water (see Tip 15). Note that some plants will thrive on this, others will hate it; this kept my spinach and chard going right through the drought, but the tomatoes turned up their toes.

Build a green urinal in your garden if you have the space. This is a tip from the National Trust in Britain, which has millions of visitors to its properties, and came up with this plan to stop half of them from flushing.

All that's needed is a bale of straw and a modesty screen. Set the straw down in a sheltered part of the garden away from any water sources. Ask feed stores or nurseries if they have any straw spoiled by mould or weeds – they may give it to you for free – or make your own bale. I compacted dead grass into a rectangle about one foot high and three feet long. Set up a waist-high screen of sticks around the straw – you can construct your own (I used discarded bamboo) or buy from a garden centre. The straw or grass deodorises the urine, the urine helps decompose the straw, and after several months, you can use it as mulch in the garden, and start again with a fresh bale.

BUILD A
GREEN
URINAL

You might already own equipment that could help in the quest to save and harvest water. Check your garage, attic or storage space for useful camping and gardening gear: camping showers, stoves and washers, garden sprayers, jerrycans, foot pumps, trailers, wheelbarrows or trolleys for moving containers of water around – all these are gold. You might have dustbins, tarpaulins, canvas tents, wheeled suitcases, cooler boxes, funnels and much more that could come in handy. And you can't have too many buckets. There should be one in every shower and next to every toilet.

Bonus tip: visit camping stores to get ideas, and draw up a wish list (see p. 102) of equipment, along with a price list, so you can plan your water budget to fit your needs and pocket. Be aware that some "dream machines" are not as ideal as they sound: air-to-water machines, for instance, are expensive, noisy, gobble up electricity and need high levels of humidity to be effective.

If you have a pool, turn it from a liability into an asset: it can become a valuable water-storing facility. Set up a system that enables as much rainwater as possible to flow into it, cover it, and use this as back-up for flushing.

Bonus tip for the future: consult an expert about turning your pool into an eco-pond that requires no chemicals to maintain. This could become a beautiful garden feature with aquatic plants and friendly frogs to catch flies and mosquitoes. Keep an area clear so the family can dip in and cool off or do a spot of water aerobics. If you want to swim lengths, plan on doing so at the gym or public pool.

WARNING! All the usual warnings about pool safety apply even more in drought conditions: thirsty animals and curious children will be more than usually attracted to water. Be 100% vigilant and make sure that your safety features are in apple-pie order.

Get inventive. Tie a funnel to the "elbow" of your satellite dish and run a pipe down from it into a container. Harvest the water from your office air-conditioner. Set up your planters to act as mini water tanks. Save catering-size containers and paint-tins.

Stock up on the following toiletries: antiperspirant (Mitchum is the one exception to my no-brand-recommendation – worth every penny), dry shampoo, leave-in hair conditioner, disinfectant, hand sanitiser, hand lotion, talc (not just for Grandma: good for no-shower days), wet wipes.

Bonus tip: on the topic of wet wipes, remember that you shouldn't flush these – NOT EVEN WHEN IT SAYS YOU CAN ON THE PACKET. Try to get biodegradable ones and put these in the compost, or make your own (there's a great recipe under **Resources**). Note that "biodegradable" and "compostable" do NOT mean flushable.

BATH-ROOMS

Heading indoors, probably the most important but also difficult areas to tackle are bathrooms, because they are so tied to our concepts of privacy and hygiene. But within the home, these are the places the middle classes use the most water, so any positive changes you make here will be of great benefit.

The first thing to tackle is your toilet. The golden rule for toilets in a water shortage: put nothing into the bowl that could contribute to a possible blockage. Not wet wipes, not tampons, not hair, not washing-up water that has oil or grease in it, not kitchen paper, not even too much toilet paper. (See Tip 36 for what to do with your pee paper.)

Find the stopcock and turn it off. Now see if you can access the cistern; if yes, this is where you should be pouring in grey water caught from the shower and the washing machine. If you fill your cistern with grey water, it means you can flush as you normally would. (This is good for guests, who may not have the knack of juggling your toilet tank lid.) If your cistern is inaccessible, simply pour buckets of grey water or other undrinkable water (dirty rainwater, for example) directly into the toilet bowl. Your aim will get better with practice!

Bonus toilet tips:
→ Before using shower water to flush, extract any hair.
→ Grey water means a grubby toilet bowl. Add bicarb, vinegar or bleach to the cistern along with the grey water, and at regular intervals to the bowl itself.
→ There's been some anxiety about possibly leaking grey water back into the clean municipal supply via toilet cisterns. With modern toilets, this is not possible, but with older models, you might like to call in a plumber to check your cistern to make sure it's absolutely safe for grey water usage.

Some clever folk are replacing their cistern/tank lids with purpose-cut wooden, plastic or polystyrene tops, often with handles attached, or small funnels, for ease of access, and also to avoid dropping and smashing heavy porcelain or ceramic lids.

You should be letting your yellow mellow (although in truth, there's no need to pee into a toilet at all – see Tip 38), but for those who sit to pee, paper should be dropped in a bin next to the toilet. You can sprinkle in some bicarb to prevent whiffs, or break a piece of incense into the bin.

Mellowing can be hard on the nose. Wee Pong (the name of this product gets a special "in-your-face" award) gets five stars from all its reviewers, works instantly and keeps the loo clean. (It's an enzyme rather than a cover-up perfume.) Similar products include G Wizz and Albex Noflush.

Bypass the toilet entirely when peeing. Millions use a bucket or potty at night and empty it outside the next morning. Those oval 1-litre yoghurt pots from Pick n Pay are the perfect shape for settling between feminine thighs, and can be emptied out into the garden.

What if you live eight storeys up, or your garden is a Zen square of raked gravel? If you have a regular bathroom, you have two outlets for (almost) water-free urine disposal: the bath plughole and the shower drain. (I was taught this trick by an elderly cat who hated litter boxes and going out on cold nights.) Pee into your container, then tip the urine down the drain, following with a splash of water mixed with bicarb or disinfectant (keep a squeezy bottle on hand for this purpose); or spritz one of the scent-killers in Tip 37. This is no more "disgusting" than having the rest of your sweat and grime and skin cells washing down that same outlet. Plus, we've all peed while showering (a sensible habit, in fact).

Put on your grown-up broeks and start researching composting toilets. These range from a bucket in a box to expensive devices with collection chambers, ventilation shafts, separators, and disabled access (the latter is an excellent idea, especially for NGOs). It seems one can bang a rustic but perfectly functional one together with some wood off-cuts and a bucket. A friend made one with a POOL NOODLE. Google is your friend (see Resources). Here we have to overcome our "ick" response; a good way to do this is to attend a talk (there are many on offer at the moment) that demonstrates how these work. Many are reassured to see for themselves that the resulting "product" is inoffensive and unstinky.

From Waterloo to Futureloo: the basic principle of composting toilets is that urine and shit need to be kept separate. Urine is sterile when it leaves the body, a valuable source of fertilising nutrients, and easy to dispose of safely. Shit is far more of a challenge: it too can become an excellent fertiliser, but it needs to be carefully handled and stored first. In its "raw" form, it constitutes a significant health risk. Interestingly, if urine is kept away from shit, the latter breaks down into a beneficial "compost" far more quickly and safely.

A bath can be an extremely useful storage depot for water caught in the shower, especially if it is close to the toilet. It's also somewhere to store other non-drinking water for indoor use: unfiltered rainwater, or spring water that isn't safe to drink. If it starts to smell stagnant, add bleach, vinegar or bicarb. To keep things fresh, instead of constantly topping it up, use up the contents, saving the last half-inch of water to clean the tub thoroughly, then start again.

If you're relying on your taps for washing up and showering, fit them all with aeration devices/filters/heads that slow down the water flow. Guesthouses drawing on municipal water should fit these immediately, if they haven't already.

Catch every drop of shower water for flushing. Buckets aren't effective enough, and can be hazardous for those not completely steady on their feet. Other options range from a child's plastic paddling pool (which can be deflated to fit the bottom of your shower) to builders' cement trays to cat litter trays. Pick something that's easy to tip into a bucket or the bath, and keep the bottom clean. Use non-slip mats if necessary.

There are plenty of options for showering (including hair-washing) that use 5 litres of water or less, and also enable you to make use of harvested water (rainwater, well water, etc).

→ Sluice buckets of heated water over yourself, or you can get a camping "bladder" shower and hoist it up in the shower or over the bath.

→ Get a 5-litre pressure-spray tank familiar to gardeners and farmers (available from nurseries, big supermarkets and hardware stores). Put in a litre of hot water, then top up with cold water until the temperature is comfortable. Pump the plunger on top vigorously, then press the nozzle at the end of the pipe and a stream of pressurised warm water will emerge. You can wet yourself all over, soap away, then rinse lavishly, and it still uses almost no water. This takes time, and you have to keep pumping, but it's effective, super-water-wise and a boon for home nursing: you could easily sit an invalid in a plastic chair in the shower, and give them an all-over rinse this way.

→ Make new bathroom routines fun, especially for kids, who will love spray gadgets, and bathing in paddling pools or buckets. If you're using municipal water, come up with games to keep showers as short as possible: for instance, counting for as long as the water is running. Get wet while counting, switch off, stop counting, lather up. Switch on, resume counting. Rinse and repeat. Time's up at 60! You could also sing songs with a set number of verses and choruses, play short bursts of music, set alarms, buzzers and beepers to go off, and so on. Imagination rules!

→ Family members could get back to the practice of sharing bathing water. Small kids can climb into a paddling pool or basin bath together. Couples could get into the shower together and take turns to lather each other up and rinse each other off. (This might have romantic benefits. See Tip 49.)

→ Or you can simply punch holes in the lid of a 5-litre water bottle, fill it with warm water and pour it over your head while standing in a basin – a beautician who surfs came up with this system; see Resources.

Kai's hair-wash method for short hair: stand over a basin and empty half a squeezy bottle full of water and diluted shampoo over dry hair. Lather, repeat if necessary, then rinse with another squeezy bottle full of clean warm water. Catch all water and save for flushing. This works well for all natural hair types.

Jane's hair-wash method for long hair: rub bicarb into wet hair and scalp. Spritz hair with apple-cider vinegar using a spray bottle. The resulting mix will fizz – treat as lather. Rinse with a squeezy bottle or jug of warm water to which you've added a few drops of essential oil, to ensure you don't smell like a chip shop. Save this water for flushing.

Good news for your electricity bill! If you're hardly ever using municipal water to shower, switch your geyser off for the duration. If you'd still like warm water to wash, and you don't want to keep boiling the kettle, here are some geyser substitutes:

→ **An urn:** good choice if you also do a lot of entertaining, or have meetings (stokvels, prayer groups, book clubs) at your home; you can heat water not only for tea or coffee, but also for the washing up afterwards, and have enough left over for a warm bucket-bath.

→ **An electric bucket** (this plugs in exactly like a kettle): cheaper than an urn, and can be used the same way.

WARNING! Handle water heated by either of these methods with great care to avoid scalds and burns, and keep them away from small children and pets. Read the safety instructions closely, and follow them.

Bathroom design has a fascinating history, and there's a lot to be learned from Japanese bathroom design, the hammams of the Middle East and Asia, and more. For instance, in some tiny Japanese homes, the bathroom sink and toilet cistern double up. One thing is clear, though: modern Western bathroom design is wasteful, unsustainable, and not always effective. Apparently showers should spray upwards, not downwards, for maximum cleansing benefit and least water use. Who knew?

PERSONAL
HYGIENE
& GROOMING

We might believe that a daily top-to-toe shower, shave and hair-wash are essential when the opposite could be true. Much of our washing culture is driven by advertising: we need all those body lotions because we bathe way more than is good for the natural health of our skin, scrubbing away at our natural protective oils.

But beyond entitlement, habit and marketing: how often DO we need to wash ourselves? Borrowing from the common sense practised by previous generations and much of the South African population, this is the mantra for daily self-cleansing: wash face, pits, bits and feet, and in that order, if we're using a bucket.

Even this isn't quite accurate: the body parts we need to clean most often? Our hands. Faces don't actually need more than a dab with a damp flannel. The skin over the rest of our bodies does not need daily washing. Twice or even once a week is in fact optimal for skin health. There is no rational basis whatsoever for our daily immersion.

Common sense suggests that if you're performing manual labour or working in dirty conditions, exercising strenuously, having a lot of sex, menstruating, working with soil or animals, or have a medical condition that requires strict cleanliness, you'll need to shower or bucket-bath more often.

TIP 45

Create water-resilient personal grooming/beauty/hygiene routines. Now is the time to try that bold new short hairdo, or grow a beard. Those with lady-gardens might consider dispensing with their foliage, given that we're all showering less. If you have a high tolerance for pain, consider embracing the full Hollywood wax. Those who use soap and water to shave their legs: switch to waxing, if you can afford it (or go gorilla).

Hairdressers and beauticians in the Cape have noticed a steady improvement in the condition of their customers' hair and skin, as they've been showering and washing their hair less often. So there is good news.

TIP 46 Use mint/tea-tree/fir/eucalyptus shower gels and soaps to feel cooler and fresher for longer. Citrus-y scents are also good, as are options like cucumber, aloe vera, rosemary and lavender.

TIP 47 If you have a bidet, you'll know they're the Rolls-Royce version of a bucket-bath. They're by far the most water-wise, convenient and comfortable form of washing that involves an indoor plumbing fixture, and a boon for the elderly, infirm, those menstruating, and before and after sex. There are very good reasons there is a bidet in every bourgeois bathroom in the southern Mediterranean – hot-weather countries in which men aren't often circumcised.

Bonus bidet tip! A bidet allows you to pee in relative physical comfort, and then use a trickle of water both to refresh your bits and wash out the basin. To keep the bidet itself spotless, pour in a cup or two of boiling water every other day, with a bit of bicarb or a splash of disinfectant if you're feeling fastidious.

TIP 48 Teen hygiene: teens, often self-conscious about their appearance, can sometimes seem almost addicted to showering, washing their hair, shaving and changing outfits. Instead of nagging and yelling, challenge them to think up water-wise ways of grooming; test-driving the best dry shampoos and leave-in conditioners, experimenting with new up-do or other "mask the flop" hairstyles and waterless skin-cleansing routines. Include them in family water-harvesting practices; standing in line at a spring with containers will teach valuable lessons about the value of the water they're hauling, and act as a visible reminder that we're all in this together.

Sex in a time of drought:

→ The saving in terms of personal washing and laundry of practising safer sex should be obvious. For long-standing couples, introducing or re-introducing condoms could make you feel like you're courting again.

→ You might not feel confident about being sexually active while showering less often. One romantic solution is for you and your partner to shower together (see Tip 43); or take it in turns to give each other a "bed bath". Use your imagination (also wet wipes, sponges, heated flannels, soft cloths, a bowl of warm scented water …) This could be a fun way to get fresh while getting fresh.

LAUNDRY, DISHWASHERS, FREEZERS & MICROWAVES

Cut down on laundry by wearing clothes until they show signs of independent life. Undies can be handwashed while showering, and if you've sweated into your exercise clothes, get under the shower in your leotard or gym kit, and wash your body and your clothes at the same time. Another option is to "air-launder" your gym clothes, or put them in the freezer for a few hours to kill the bacteria that cause smells.

Tablecloths can be retired or replaced by the wipe-clean kind. Wear a plastic apron when cooking.

If you have a washing machine, check which programme uses the least water. In many makes, it's not the economy or the speed wash, but the synthetics wash. Running the washing machine uses a LOT of water (anywhere between 40 and 70 litres per wash, yikes), so try to collect it all for flushing.

Bonus tip: when putting your washing machine outlet pipe in a bucket, tie or weight that sucker down. They take on a life of their own. Also make sure your container is big enough to hold all the water: as much as 70 litres can gush out.

Another option is to rig up your washing machine outlet pipe to a dedicated container. If you're really organised, set up a pump that will deliver this water to your toilet cisterns. Whether you are able to do this will depend on the layout of your home, laundry and bathroom. If this is possible, it should be a permanent installation.

TIP 53

Replace your machine with a low-water and electricity-free camping washer, known as a Sputnik, which has become so popular overnight, there's a waiting list to buy it.

If you can't wait, you could try the Fiona-Patsy™ cooler-box method.

Get a hard-sided, sealable cooler-box. Heat a few litres of clean but non-potable water and pour it in with about another 2 litres of cold water. Then add detergent: I make a soup of biodegradable washing powder, bicarb and Vanish Liquid. Don't use your regular washing-machine powder! Choose a soap/detergent that doesn't foam. Woolite or any hand-wash laundry soap works well. And lose the fabric conditioner – it's unnecessary for line-dried clothes, and bad for the planet. Next, sort the laundry as usual, pre-treat stains, and toss everything in. Don't fill the box to the top. Stir it all up with a wooden spoon.

Next, seal the lid carefully (this is vital) and shake the box for a few energetic minutes. If you have errands to run and a car, wedge the box in the footwell and head out. (The slosh-slosh sound as one corners is rather comforting.) An hour or two later, pour the water out into the grey-water cache and add 3 litres of non-drinking water back into the box. Repeat the cardio workout and drain the water out again; if it still looks murky, rinse one more time with another 3 litres (taking total H_2O usage to around 12 litres, all kept for flushing). You can now hang it out (I remove the "DECOMMISSIONED" sign from my washing machine, chuck all the wet clothes in, and hit the "Spin" button).

Other cheatipants ways to do the laundry: if good rain is forecast, pre-soak your dirty clothes and spot-treat and scrub stains. Then hang everything on the washing line and cross fingers for a good soaking shower or at least an hour's drizzle to do the rest. This will also require a day of bright sunshine for drying afterwards, so it's a bit of a gamble, but very satisfying. I've also "washed" towels simply by pegging them to the line when the rain starts.

One more popular option is to soak and treat for stains, then dump everything, including the soaking water, in your machine, and wash using only the rinse cycle. This uses only about 20 litres, all of which you can save for flushing. If you have plenty of clean but undrinkable water, you can even run this water in as the rinse cycle starts; with front loaders, it's a bit tricky as you have to pour it in via the soap dispenser, but it's do-able.

Sally's cute laundry tip: once you've handwashed your undies, use a salad spinner to rid them of excess water.

Menopause without regular showers is no fun. If you battle with constantly damp nightwear and sheets courtesy of hot flushes, hang everything out on the line each morning to air-dry. It's not a perfect solution, but it does freshen things. And give yourself plenty of cool-water sponge-baths.

DISHWASHERS, FREEZERS AND MICROWAVES

TIP 58

If you're reliant on municipal water for washing up, and you have one, a dishwasher uses less water than washing up by hand. Treat it as a storage place for dirty dishes and don't run it until it's absolutely full. If you run it at a high temperature (60 degrees), this should kill bugs left by Fido's tongue (although human mouths are apparently filthier than those of most animals). But if you have access to clean non-potable water, use this to wash up by hand, and mothball your dishwasher.

Bonus tip: dishwashers use less water than washing by hand ONLY if dirty crockery isn't rinsed first. Use the lick, scrape and teabag tips before filling the dishwasher (see Tips 62 and 64).

If you have a chest freezer, use your big pots once a week to cook large batches of rice, pasta, soups, dhal, stews, pap, curries and any other dishes that need plentiful water for cooking. Freeze meal portions in microwaveable containers. Use the teabag (Tip 62) and napkin tricks for cleaning the pots afterwards, and if you're lucky, you could even set them out to soak when it rains.

Water-wise ways with your microwave:

→ Cook or reheat as much as possible in your microwave. Almost all veggies can be cooked in the microwave with a scant spoonful of water and maybe a bit of butter added – almost no water, and fewer pots to wash.

→ **Sue's super-bonus-hygiene tip:** if your knickers or underpants are PLAIN COTTON (no bits of synthetic lace and ribbon or metal studs), after washing them and spinning them (see Tip 56), you can put them in the microwave for 30 seconds at a time. This will both dry them and sterilise them. Watch carefully; you do not want smouldering knickers. Well, not literally.

→ Facecloths and flannels: dampen these and then heat them in the microwave before using (the same principle as those comforting hot cloths they hand out on airplanes before meals). This has the added bonus of sterilising these cloths. It's also something nice to offer guests before or after serving food, especially if folk are using their fingers to eat.

MICROWAVES
ARE GREAT FOR ALMOST
ALL WATERLESS COOKING.

KITCHENS, COOKING & FOOD

The water crisis is a chance to rethink kitchens and cooking. I love cooking for friends, and have no intention of giving up parties and celebrations. So how can we change our kitchens, our recipes and our entertaining habits so that we can go on having a life, spending time with friends and family, without feeling like collective water Grinches?

TIP 61

The biggest water-wise tip in the kitchen? Don't. Waste. Food. Ever. All food takes an astonishing amount of water to grow, clean, package and process. Leaving one egg uneaten on your plate represents a staggering loss of 196 litres of water (yes!!), according to the Institute of Mechanical Engineers (see Resources).

Bonus tip: this also applies to food in restaurants. Ask for a doggie bag or take Tupperware along and insist that leftovers go into these to take home. (Do this just once, and your teen will never again take a single bite of a burger and then abandon it.)

DON'T. WASTE. FOOD. EVER.

Check out the lists of common foodstuffs and how much water each takes to produce on the Resources page. You'll immediately see that meat takes a truly spectacular amount of water to produce, pound for pound. This shows why an important long-term water-saving habit is to eat less meat, or to adopt a vegetarian or vegan lifestyle. The good news is that tea, wine and beer take the least amount of water to grow and process. So put the kettle on.

TIP 62 **The teabag tip at last, with thanks to the Teabag Elf:** "I keep all used teabags to wipe dregs, oil, foodstuffs from plates, forks, pots, peanut butter bottles ... many kitchen items can be cleaned by wiping all food gunk off with one or two damp used teabags and then drying with one or two dried used teabags. I'm uncertain about any bacterial health hazard but I have been doing it for a couple years now and (I think) I'm not ill or dead. I sometimes re-use the cleaned item (same plate and cutlery for several meals) or else let these wiped dishes pile up unstinkily and have a marathon dish-wash in as little water as possible."

My friends in catering have all confirmed that this is by far the best method to get most grease off pots, pans and plates, also to mop up cooking-oil spills. And the teabags can go straight into the compost tub afterwards.

TIP 63 **If you don't already have a compost bucket in** your kitchen for fruit and veg peelings, set one up now: this is where to scrape plates after meals. Empty this bucket regularly and sterilise it with bicarb, vinegar or Milton once a week.

TIP 64 **Eat directly out of pots, pans and containers.** Allocate each family member a mug, glass, plate (for dry food, like toast) that they have to keep going as long as possible before washing. If a plate has to be washed, lick it first (the kids will love this, and the family dog will be glad to help out). Do be sure to wash at a high temperature afterwards (see Tip 58).

EAT
DIRECTLY
OUT OF
POTS

Have more braais. Get your potjie into the action. Cook your food on the coals, then burn the paper plates and napkins afterwards.

Retire your goblet blender. It takes way too much water to clean. Replace with a stick blender, which I use directly in the pot or whatever container I'm going to be eating from. Proceed with caution: you want something narrow and deep to avoid spatter.

Retire your big pots. Almost all my cooking is now done in non-stick frying-pans and saucepans. Try to get ones with deep, steep sides. The magazines all say "get the best one you can afford", but I'd rather recommend that you don't buy the cheapest, in case the lining peels off the first time you use it. Mine usually just get wiped after use and washed about every third time they're used for cooking.

Remember the wok craze? If you have one, dig it out and treat it as a non-stick frying-pan with high sides. It should be possible to clean all these utensils with teabags, kitchen paper and only then, if needed, a little splash of water and dishwashing liquid.

TIP 69 **Your partner in non-stick cooking and waterless clean-up** is a whole bunch of silicone spatulas. I got given an indestructible one from Le Creuset, and it gets used to stir everything during cooking, to decant food out the pot for serving, to get the pan as clean as possible, and to scrape the plates into the compost tub afterwards. I have several cheap plastic ones as well, to make sure that containers have every bit of food residue cleaned out before washing.

TIP 70 **Hunt down the best one-pot meal recipes:** the days of a separate dish for cooking meat and three pots for the veggies, plus a gravy-boat, are done. There are wonderful recipes for meals that combine protein, veggies and starch all in one pot: I typed "one pot meals" into Google and got "about 81 300 000 results", so you are never going to run out of ideas. Many of these are also budget meals.

TIP 71 **Eat more raw food and finger-food.**

TIP 72 **If you have no objections to meat or alcohol,** here's Megan's deglazing idea – add a splash of wine to fatty pans in which you've fried meat, and then gently rub the bottom with a spatula. Reserve the resulting liquid for the next sauce you make.

TIP 73 **Sticky bottoms:** if you do have to wash a pan and the bottom is sticky, sprinkle a little of your magical all-purpose Swiss Army bicarb on the bottom, add the merest splash of water and let it stand for a while before giving it a gentle scrub. Lifts off even burned-on milk or egg residue with no fuss.

Ditch the crockery. All over Indonesia and in many other tropical places, banana leaves are used as plates for even wet, soupy dishes – straight from the table to the compost heap afterwards. There must be a business opportunity here (eyes the KZN coast).

In the absence of leaf plates (not everyone will be as enchanted by this idea as I am), here's another way to serve guests food, especially at parties: bamboo "banana boats". Much sturdier and easier to handle than paper plates, they work for sit-down and stand-up eating, hold a good portion of even wet and messy foods without leakage, the sides prevent your falafel from leaping to freedom, and they can go straight into the compost heap at clear-up time. See **Resources** for where to purchase (you can get them online).

Compostable coffee-cups are a good option for workplaces where cups stack up in the communal sink. Take these home to dig into the garden or burn. Add a few to your out-and-about basket to use in place of glasses.

Related to Tip 76 is the issue of garbage: if you're used to composting and recycling, you need to head guests off before they start hunting for your rubbish bin. When they offer to help clear up, ask them to scrape their plates into the compost tub.

TIP 78 **A consideration for well-meaning guests:** if your host has served you dinner on regular crockery, and you leap up and start stacking dirty dishes, someone will have to scrape and wash both sides. Carry plates into the kitchen one by one instead. If space is limited, simply stand by and offer chocolates.

TIP 79 **When having friends round,** put out a basin of warm water with some fragrant bubbly handwash added. You could use an old-fashioned jug and ewer set, and pretty handtowels. You'll find that people will put down the salad or cake or whatever they've brought, and immediately look round for somewhere to wash or wipe their hands. This also means everyone can wash their hands before eating or after using the toilet without going near a tap. Once this water is cold and the bubbles gone: more grey water for flushing.

This could be an opportunity for members of your family to work more closely together preparing meals, discovering new recipes, and having friends round more often. If you're water-resilient, your friends who are dependent entirely on municipal water will be especially delighted to get invitations to a meal or pot-luck dinner.

POT-LUCK

COMMUNITIES & NEIGHBOURS

During apartheid years, street committees supported individuals and organised action before cell phones and emails came along. Capetonians are rustling up those skills all over again. The basic principles involve lowering the drawbridge as opposed to retreating behind our walls.

We need to rediscover the meaning of neighbourliness, to feel a sense of connection and community. Given our awful and still-painfully recent history, this is not easy. But we're in this together, and we need to give each other all the boosting we can, as well as sharing resources – including the ones between our ears – wherever we can.

TIP 80

The Golden Rule: be a good neighbour. In reports on the water crisis in Brazil's São Paulo (see Tip 1), some commentators noted that a creeping disaster turned people against each other, as opposed to a sudden disaster (tsunami, earthquake, etc), which tends to pull people together. And this is related to the next tip.

TIP 81

Accept that blame is pointless. Yes, it's infuriating watching politicians and municipalities supposedly responsible for our welfare stick their heads in the sand, exploit the crisis for vote-grubbing reasons, and blunder from one PR gaffe to another. But there are better uses of time and energy than fuming with helpless rage.

Sadly, Day Zero scenarios present opportunities for criminals: water theft is on the rise, and there are those who will prey on the weak and the confused. Let's say Auntie Mavis, who has asthma and no car, is thirsty and desperate – and opens her door to see a likely-looking lad flashing a name-badge and offering to collect water for her for a hundred bucks. She hands over the cash, and never sees any water or her money again. So get to know Auntie Mavis and establish trust now so that when the paw-paw hits the fan, you're a familiar face.

TIP 82

Consider whether you're prepared to share. One suggestion has been that everyone with an unstressed well, wellpoint or borehole identify three "adoptees": a neighbour; a nearby vulnerable family; a nearby vulnerable NGO. And then commit to giving them water for flushing and basic hygiene once Day Zero arrives. If you have a source of groundwater that is plentiful and perennial, put this plan into action before Day Zero, in the interests of eking out the last few puddles in our dams.

TIP 83

If you have a car and are able-bodied, every time you collect water from a spring, take a few folk who can't manage on their own. If you own a bakkie, load up as many people and their containers as you can, and get the huskier folks to do the lugging for the frailer ones. Make this a regular gig. Separate your teens from their electronic gadgets, give them sunscreen, hats and reflective vests, and send them to the nearest spring for an hour or two to help people carry water.

TIP 84

Plug into existing community networks – churches, mosques, synagogues, all those parental mafia groups that circulate around schools, clubs, charities, and see which can be adapted into "water webs". Don't interfere, gossip or stoke existing tensions, but do ask people what plans they have, and if there is anything you can do to support them. (If nothing else, help them get online and show them how to find reliable water information.)

Neighbourhood watch networks can be adapted with great success. An extremely successful neighbourhood support/watch system, which has been able to switch effortlessly into a water network, has the following to add to the above tips.

→ Establish where the geographical boundaries lie, so as to eliminate gaps and overlaps. Meet with others in your suburb or complex or estate with maps, if necessary. A block of about 300 households is manageable. Make sure everyone at every address is accounted for.

→ Pick one central means of communication. Email is a bit laborious, but it's better for the elderly, who aren't necessarily on top of social media. Email is also less likely to generate trollery, racism, conspiracy theories and other nasties, as people have time to think before reacting, and their names are attached to their mails.

→ Start the old-fashioned way: a letter delivered to every home. A few volunteers will need to do the admin in collecting everyone's email addresses and cell numbers and generating a comprehensive list.

PLUG INTO EXISTING COMMUNITY NETWORKS

→ Ask people to share information about their water resources/resilience. You may find that if everyone is super-thrifty about water and generous about sharing it (especially in wealthier or less built-up areas, where there's more room to install tanks and swimming pools are plentiful), you might not have to queue for water come Day Zero.

→ But if you do, establish networks for that as well. Who has transport? Who has the physical ability to carry? Who works flexitime or is their own boss? Who would be prepared to collect water for others? Don't guilt or shout anyone into agreeing to do this: rather expect to be pleasantly surprised.

→ You might need to form a committee to deal with awkward and tense situations: Mr Twitwiddle has been sneaking his hosepipe into his pool; Dr Leakey has been spiriting away water from the tank installed at her block of flats to keep her roses going, and so on.

Bonus tip: take cake to all potentially tricky committee meetings. I know carbs are supposed to be the enemy, but they definitely have a tranquilising effect.

→ To quote the organiser who supplied these tips: "Our community is on a path to a shared solution and one that brings people together to care for the most at need and then the collective – very cool outcome for a crisis."

**DON'T
SHARE
FAKE
NEWS**

TIP 85

Don't share fake news via social media, something all too common in an atmosphere of panic. Never share a link on the basis of the headline alone. Never share a report that doesn't include the actual original documents supporting the claim.

TIP 86

Don't get angry or self-righteous about people who need water for their animals. Every time this comes up on social media, there's indignation about "caring more for animals than humans". What is one supposed to do when pets or livestock are experiencing thirst and dehydration? Say, "Sorry, we were selfish and stupid not to plan for this, and now you're going to have to suffer on principle, because, you know, humans are more important"? Likewise, putting out a tiny bit of your water allowance each day for birds and insects is not a hippy-dippy luxury. No birds and bees = no pollinators = NO FOOD.

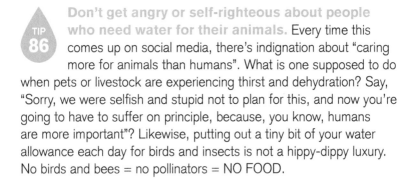

TIP 87

And speaking of food chains, it doesn't make sense to complain about water use by agriculture, as if growing food was some sort of optional extra. Almost everything on your plate started its life in a farmer's field, no matter how much it's been processed and packaged since then.

This crisis raises the age-old problem of human selfishness: how do we tackle it in the face of a serious problem that requires concerted communal action? Psychologists suggest that strategies for getting people to change their behaviour fall roughly into three categories: shaming them; making the change cool; and rewarding them. So: sticks, basic advertising and carrots. Education and inspiration are also vital tools.

Shaming is tricky territory, and some earnest conversations may be needed. People are rarely truly evil; it's more likely that they haven't really imagined the consequences of their behaviour.

Then there's advertising: modelling "best" behaviour so that it becomes the in thing. In London, for instance, no self-respecting hipster would dream of asking for a takeaway cup in a coffee-shop: all the cool kids bring in their own travel mugs and glass juice jars.

Finally, carrots: offer rewards – in your home, business, school and university – for being the best water warriors. Praise and acknowledge efforts by individuals and groups.

SHAMING
IS TRICKY
TERRITORY

BUSINESSES, INSTITUTIONS & EMPLOYERS

The water crisis has already cost hundreds of thousands of jobs, often in sectors that employ people without extra resources or capital. So the first rule of business should be to keep workers employed, if at all possible. If we reach Day Zero, offices might have to close, and schedules are going to be thrown out by the need to collect water; businesses should prepare for these eventualities even as we hope they never come to pass.

One ray of hope is that businesses offering flexitime and "work from home" services are going to be better equipped to hang in there and even thrive. In the meantime, here are some tips for businesses both large and small.

If you have offices around the country, and high-skill employees with flexible domestic circumstances, find out if they would be willing to relocate to another city or become weekly commuters for the duration. Wherever possible and suitable, allow employees to work from home.

Appoint a water task-team at work. Your existing Health and Safety representatives may be the best people for the job. Draw up guidelines for use of communal bathrooms that strike a balance between water-parsimony and dignity (putting a paper "clock" or tear-sheet on the doors of toilet stalls indicating that flushing can take place every third wee, for example; switching off water to sinks and setting out hand sanitiser and wet wipes, and so on).

This is probably a good opportunity to check your fire protocol and run a fire drill as well. You may need to stockpile both bottled water and non-potable water; the cleaning staff will be able to do their jobs, and flushing will be a less tricky issue if there is a store of rainwater or similar on or nearby the premises.

If you share office premises with multiple businesses, ask your water task-team to reach out to their peers and form a building committee. This is especially necessary if toilets are communal, or if you need to purchase shared water tanks and install rain-harvesting systems. Set up meetings with the landlords as soon as possible to discuss responses to the water crisis. As with your own office, the janitorial supply cupboard should be stocked with non-potable water for cleaning and flushing emergencies. Set up a system for harvesting and hauling this water, and ensure that those who do so are fit and recompensed for their time and any transport costs.

Talk to staff about the impact of the water crisis on their routines. Be aware that in many informal settlements, water pressure to standpipes has already been reduced, and collecting water takes much longer. Find out if employees need time off to fetch water, and discuss rosters in case of Day Zero.

Schools and similar institutions should set up water task-teams along the same lines as businesses, with parents and staff teaming up. Keeping schools open is critical, and water storage and harvesting processes must be set in place urgently. Fundraising may be necessary to install wells or boreholes, and if space allows, set up waterless urinals as swiftly as possible (see Tip 28).

If you employ a domestic worker in your home, do not burden them with extra tasks brought on by the water crisis – they are no doubt already expert at using as little as possible when doing their own household chores. Rather consider re-allocating tasks. If practical, do some of your own chores: dishes, laundry, toilets, basins and floors (which we wash far too often, anyway). Your employee could focus instead on sweeping, dusting, polishing, vacuuming, ironing and food preparation.

Now is the time to relax our standards of "cleanliness". Dust doesn't hurt, and spiderwebs are great for catching flies. Academic Lesley Green points out that we confuse "shiny" with "clean", and will consider surfaces wiped with a film of chemicals to be "clean" when in fact they simply look that way. Rather give your family the message that the responsible and considerate way to proceed is to be **tidy**: pick up after yourself, put the milk back in the fridge, and hang up that towel.

CONSIDER TIME-SHARING

If your business is water-resilient because you have a well or borehole, or access to either, consider time-sharing temporarily with colleagues dependent on municipal water, who might have to close their premises because of Day Zero. This is especially true for businesses that rely on water: hairdressers, laundromats or florists. One set of staff could operate from 9-3, then have colleagues come in from 3-9, and so on. Laundromats with independent water supplies could offer a 24-hour service – something many will need as the dry months drag on.

TIP 94

It may be time to call a halt on giving cut flowers as gifts. We all love the beauty of flowers, but florists might want to stick to fruit baskets, pot-plants and dried flower arrangements for the duration. Instead of sending flowers for bereavements and births, why not send food hampers and baby packs?

TIP 95

→ One excellent water-wise idea already in practice at some Cape schools is to let pupils come to school in gym clothes (appropriate where children have PE outfits – not every family has the budget for this). Another option would be to allow pupils to wear T-shirts instead of white shirts that need daily washing.

→ Likewise, every industry that requires staff to dress in a professional uniform for reasons other than safety (banks, call centres, and so on) should issue staff with branded T-shirts to replace white shirts, and, once it gets colder, tracksuits to replace professional outfits that need regular washing and ironing. If yours is a small business, consider making it "Casual Friday" every day.

→ The above two tips might take a bit of planning, but something every business and school could and should announce immediately is, unless you need closed shoes for safety reasons, allow anyone who wants to wear sandals or flip-flops to work. No more sock-washing!

→ The Navy could drop white dress uniforms for the duration, and wear camo instead.

TIP 97 Universities should issue all students in residences with buckets or pressure-sprayers, and a pack of wet wipes (but everyone repeat, DO NOT FLUSH THEM, EVER – blocked toilets are never fun, less so in a drought, and least of all when they're communal).

Bonus tip: universities and colleges might like to challenge their residence students to compete to use the least water per person per month, with the winning residence getting a pizza party.

RESTAURANTS

A Cape Town chef and restaurant owner sent all her regular patrons this sensible and reassuring letter, with such good tips that I've repeated it here, with her permission: "With the all-consuming thoughts on everyone's minds on how to exist on 50 litres per person per day and how we can avoid Day Zero, we have done some real head-scratching on how our restaurant can contribute to saving water.

"We have decided to reduce our water usage by serving our meals on biodegradable, compostable plates, made from sugarcane fibres. They are water- and oil-resistant, hygienic and completely compostable within four weeks, so they will be taken home each evening, to be dug into the garden, thus assisting with mulching and enriching the soil.

"This is not compulsory – we will offer our patrons the choice – traditional plates or disposable! If this step is supported by our community, we will bring in compostable cups for coffee and cold drinks. "We will not be offering municipal water at tables, but have reduced the price of our bottled water.

"We take care to ensure that our water is not bottled in drought-stricken areas – our favourite coming all the way from KwaZulu-Natal, where water is abundant.

"We will also restrict the usage of ice, which we make ourselves from bottled water. All our drinks are refrigerated and ice will only be served if requested. We have invested in wine skins to replace ice buckets, or we offer to keep your wine in our fridges and top up your glasses as required.

"We will also put buckets of grey water in the toilets, which can be used for flushing. We are testing and supplying a hygienic, waterless 'handwash'.

"We ask patrons to assist us in saving water by using these alternatives, but point out that this is purely a request – not compulsory at this stage."

The Great Tap-water Debate:
This falls squarely into
the ambit of Middle-Class
Problems, but it's roused a
surprising amount of strong
feeling. On the one hand,
waitrons say their single
biggest gripe right now is
customers who ask for glasses
of tap water with ice, and then
fail to drink more than a sip
or two. On the other, many are
irritated when restaurants ban
the serving of free tap water,
and insist that their patrons
order bottled water.

Restaurants, pubs and
coffee-shops: your clientele are
feeling scratchy, and they've
just put rain-tanks on their
overdrafts and credit cards. So
don't gouge, or appear to be
exploiting a genuine emergency
in order to gouge (this goes for
all industries). If you genuinely
no longer want to serve tap
water, tell folk they're welcome
to bring their own flask and ice.

GUESTS,
VISITORS & TOURISTS

Thinking of visiting Cape Town for work, a conference, or to see family and friends? Planning a tourist trip here, or to other cities that are teetering on the verge of water shortages, such as many Californian urban regions or Barcelona?

First of all, we want you here. We love our friends and family, and our tourism industry (which creates thousands of jobs) needs you here (although the next few months might not be the ideal time for a relaxing beach or pool holiday).

Next, while you're here, note that Cape Town's level 6B water restrictions, of using no more than 50 litres of municipal water per person per day, apply to every single person in the city, visitor or not, unless specifically exempted by law (patients in hospitals, for instance). Some Cape Town guesthouses, terrified of alienating paying guests, are relaxing restrictions or telling visitors they "don't apply". Water restrictions are NOT optional extras that your B&B has a special licence to ignore. Check if their water comes from a borehole or municipal sources before taking a long shower.

Minimise your water-guzzling laundry requirements. Guesthouses, hotels and Airbnbs could offer a small discount or reward to visitors who bring their own linen and take it away again to wash at home. Flat sheets and sleep sacks (see Tip 101) take up little suitcase room; and even if you're coming with just hand luggage, you can fit in a pillow-case.

Plan your packing so that you won't need laundry facilities while here. Regular travellers will already know that pantyliners prolong underwear life.

Shower and wash your hair before getting on the plane; pack hand sanitiser, dry shampoo and a travel mug.

Visitors from out of town to private homes can also bring their own bed linen and take it home again. Some of you may remember the sleep sacks of youth-hostelling years: turn to **Resources** to find out how to make one.

... AND YOU THOUGHT YOU WERE GETTING 101 TIPS ...

If staying in someone's home, check with them before showering, and then remember that the water shouldn't run for more than 90 seconds: basically, short bursts which you use to get wet and rinse. Forget showering every day.

Don't generate piles of washing-up, or leave food on your plate. If you offer to wash up, find out what the protocol of the house is before merrily running a sink full of hot water. Remember that it is a sin to pull a plug absent-mindedly.

Residents of drought-stricken areas should also be good guests. Prepare a pack to take when visiting friends or meeting them in coffee-shops: this could include a travel mug for coffee (quite a few places offer a discount if you bring your own mug); your own water and ice in a thermos; and I like to take a 5-litre bottle of non-potable water. This last-mentioned item is because it seems a bit over-familiar to let my yellow mellow in someone else's loo, plus what if my paper is the proverbial last straw that blocks my host/ess's plumbing? If you're water-wealthy, leave this behind in the restaurant toilet. Someone will be grateful.

THE FUTURE

If we all pull together, we may be able to keep postponing Day Zero, and even push it back into the medium-term future. However, the demand for water is going to increase even as the supply dwindles in coming years and decades. Tempting as it may be to sink back into a sense of security when (and if) the dams fill up, we need to be proactive now about how to prevent the present situation revisiting us again and again in the future.

This will require concerted civic and political effort.

We should have started pushing hard for rebates, subsidies and tax credits for installing water tanks and composting toilets decades ago. It makes no sense that there are tax incentives for installing solar panels, but not for water-storage tanks or grey-water recycling systems. It's not too late: put strong pressure on local, provincial and national government for legislative change and tax amendments, and make it clear that you will vote accordingly.

We need to tackle red tape and cumbersome bureaucratic regulations that make it difficult to respond appropriately to civil emergencies.

→ Many institutions that need to take urgent action to save water are insisting on snails' pace purchase and procurement processes. By all means, follow steps that ensure transparency and honesty, but a city running out of water and provinces classified as national disaster zones all count as major emergencies.

→ Private retirement villages, golf estates, gated communities and security estates have on the whole dragged their feet woefully on water security. New houses are being constructed with built-in guttering that makes water harvesting all but impossible, and "it's against the rules" to alter these. Dozens of criteria must be met before tanks may be installed – unbelievably, most of these relate to "appearance". (I have yet to see a garage that's attractive to look at, but a water tank is apparently an eye-sore.) All complexes, especially those homing the elderly, should radically reform and adapt their "variance" regulations to enable residents to become water-secure as soon as possible.

We need changes in building legislation. It should become compulsory for new dwellings and offices to be constructed with water-harvesting and recycling systems in place. Push hard for these.

We need to revamp basic Western bathroom design, from scratch. En-suite bathrooms are among the worst (and most unhygienic) offenders: it makes no sense to put the source of the most moisture in the home feet away from the place with the most soft furnishings. Duvets, mattresses and billowing steam are not a good mix. Bedrooms with attached bathrooms belong in guesthouses and hotels, not private homes.

The average two-or-three bedroomed home should revert to the system of one bathroom, which contains a bath (optional), shower and sink. There should then be a toilet with a sink for handwashing in a separate room. A bidet in one or both these rooms would be a damn fine idea. (This is the bathroom design for most middle-class homes in Europe.) And no more spa baths. That's what spas are for.

Don't cover your soil. Concrete, tarmac, paving stones and decks cover the earth, and direct rain run-off into drains and the sea. The earth is a valuable sponge that helps hold our water supplies.

Go indigenous in your garden. We need to stop creating thirsty exotic gardens, dig up our lawns and replace them with beautiful, hardy, indigenous groundcovers and plants.

Interestingly enough, the only cure for global warming is to generate more water – by planting indigenous trees and shrubs. These release moisture into the air, hold it in the soil, and have a moderating effect on temperature. If you want to leave a truly valuable legacy, plant a small forest.

DONATE AND RECYCLE

→ Almost everyone is being hammered by the recession, budgets are cut to the bone, businesses are struggling, but: if you can, select a local NGO and pledge to make a water-saving contribution for the next several months. Pick one small enough for your contribution to make a real difference. Talk to their staff to find out what they need.

→ Bottled water is the one thing that NGOs, soup kitchens and schools need:

→ The Dis-Chem pharmacy chain makes it easy for you to donate simply by adding the price of your water gift to your regular purchase; or you can do this online.

→ The Gift of the Givers Foundation is taking a broad approach to the water crisis: drilling wells, providing food parcels to farm workers and fodder to livestock, donating tanks and bottled water: contact them (see the Resources list) or a similar charity to find out how you can help.

→ The Water Institute of South Africa has created a platform whereby anyone can volunteer (time, money, goods) to help with the water crisis. You can find an online form to fill via the Resources on pp100–101.

→ Shoprite allows shoppers to make a small donation to the Western Cape Disaster Relief Fund (ear-marked for the water crisis) at any till. Other businesses and chains are sure to follow suit.

→ So you've installed tanks and they've filled up! Scoop up the overflow the next time it rains, and drop this water off at your nearby elderly neighbour, vulnerable family, small NGO or animal shelter. Explain that it's for flushing and washing, not drinking (although animals will be able to drink it).

→ When donating water, indicate clearly on the container whether it's suitable for drinking or not. Avoid situations where recipients might drink non-potable water, or use drinking water for flushing.

→ Spare a thought for those who have to go on washing cloth nappies in the coming months because they cannot afford disposable nappies. If EVER there was a time to donate disposables to NGOs and crèches caring for babies and toddlers, it's now.

→ Many charities and feeding schemes have had to switch to recipes and meal ideas that generate as little washing up during prep and after the meal as possible. One sad result is that many hungry children who rely on school feeding systems for their one hot meal of the day are now getting a sandwich instead, as rice, pap, soup, stews and soya mince take a lot of water to cook, plus there's the difficulty of cleaning the pots afterwards. You might like to "adopt" a local school or soup kitchen, and commit to providing them with bottled water for cooking and/or non-potable water for washing up.

→ The PET plastic bottles in which water is sold MUST be re-used and/or recycled. Every part of them is recyclable, and it's safe to re-use them – they are BPA free. So never chuck these out with the garbage; recycling centres and the manufacturers are eager to repurpose them, and to turn them into products like clothing, insulation and bedding. There are also projects that use them in the manufacture of owl and bat boxes (see Resources).

RESOURCES

Bamboo "banana boat" plates and cups: these can be purchased from Merrypak or Green Home, as can compostable cups, bowls and containers. http://www.merrypak.co.za/ or https://greenhome.co.za/

Composting toilets: https://docs.google.com/file/d/0BzVCEQx514wILU9JdnAtcmhpV0k/edit

Donating: how to give water, goods, or volunteer your time:
→ **Gift of the Givers Foundation:** http://www.giftofthegivers.org/disaster-relief/south-africa/1165-2018-disaster-relief-sa/cape-town-drought/7008-cape-town-water-crisis
→ **Water Institute of South Africa:** http://wisa.org.za/2018/01/25/water-institute-launches-platform-for-day-zero-volunteers/

Food wastage and water use:
https://www.theguardian.com/news/datablog/2013/jan/10/how-much-water-food-production-waste
https://www.watercalculator.org/water-use/water-in-your-food/foods-big-water-footprint/

Grey water: http://resource.capetown.gov.za/documentcentre/Documents/Graphics%20and%20educational%20material/Alternative_Water_Resources_Greywater_English.pdf

Japanese bathroom design: https://japantoday.com/category/features/lifestyle/12-awesome-features-of-japanese-bathrooms-you-wont-find-in-the-west

Plastic water bottles – how to recycle these: http://petco.co.za/safely-reuse-recycle-pet-plastic-water-bottles-drought/

Showering after surfing or swimming – the beautician's method: https://missmelissawrites.com/2018/01/20/how-to-shower-and-rinse-your-wetsuit-with-only-5l-of-water/

Sleep sacks/sheets – how to make your own: http://www.trusty-travel-tips.com/travel-sheet.html

Western bathroom design: https://www.theguardian.com/lifeandstyle/2014/jul/15/why-modern-bathroom-wasteful-unhealthy-design

Wet wipes – how to make your own biodegradable ones: https://www.thekitchn.com/how-to-make-wet-wipes-from-paper-towels-229046

World Wildlife Fund water tips: http://www.wwf.org.za/our_work/water/

SHOPPING/WISH LIST

→ Rainwater tanks

→ "Hippo" roller

→ Water bladders

→ Dustbins with fitted lids

→ Water containers of various sizes

→ Funnels and pipes

→ Flexipipes (also known as gutter sleeves)

→ Rain chains

→ Planters

→ Buckets

→ Basins

→ Jug and ewer sets

→ Spray bottles

→ Squeezy bottles

→ Tarpaulins

→ Wheeled trolleys

- Wheelbarrows

- Pull-along wheeled suitcases

- Camping stoves

- Camping showers

- Camping washers ("Sputniks")

- Cooler-boxes

- No-foam laundry detergents

- Garden pressure sprayers (for showering)

- Shallow, flat containers for catching shower water (inflatable children's pools, cat litter trays, builders' cement trays)

- Basin baths

- Non-slip bathroom mats

- Compostable plates, dishes, bowls and cups

- Travel mugs

- Re-usable chopsticks

- Tummy/bladder first-aid kit (over-the-counter medicine and treatments for nausea, vomiting and diarrhoea and bladder infections)

- → Toiletries (dry shampoo, leave-in conditioner, talc, hand sanitiser, anti-perspirant)

- → Fragrance oils

- → Wet wipes

- → Disposable nappies

- → Composting toilet

- → Straw bale and outdoor screen for constructing dry urinal

- → Wee Pong, Albex Noflush and similar toilet odour neutralisers

- → Small bathroom bins

- → Sterilising fluids and disinfectants

- → Bicarbonate of soda

- → Spirit vinegar

- → Thick and thin bleach

- → Water-purifying tablets

- → Compost tub

- → Water filter with ceramic "candles"

- → Water filtering jug

→ Electric buckets

→ Urns

→ Stick blender

→ Heavy-duty, steep-sided non-stick pans
   or woks

→ Plastic or silicone spatulas

→ Braai equipment and potjie

→ Gas stove

ACKNOWLEDGEMENTS

Writing this book, at high speed, took much more than a village. When I began a blog on water-wise ways the day after the city administrators finally admitted, in late January, that Cape Town was about to run out of water, literally hundreds of people wrote to me with ideas. I am touched by and grateful for your enthusiasm and determination to make a difference. I wish I could name you all – but some of you were anonymous, many offered duplicate tips – and your ideas keep coming. Don't stop!

I've tried to list everyone whose tip or actual words appear in this book, but I've no doubt left some of you out. If I've used your tip, and omitted to thank you, please let me or the publishers know so we can credit you in future editions.

Something I've loved about this unexpected project has been the chance to talk to so many people from different walks of life. To all the cleaners, beauticians, waitrons, psychologists, social workers, gardeners, small business owners, doctors, physiotherapists, hairdressers, cooks, green activists, guesthouse staff, scientists, academics, administrators, councillors, writers, journalists and strangers in coffee-shops and at springs who've shared their insights and experiences with me: thank you.

Special thanks go to: Cheryl-Anne Carrick and the staff at Oshun Hair & Beauty Salon; the staff at The Daily Coffee Cafe, and the Foodbarn Deli, especially Edward Mandikisi; Liesel Coetzer and the Simon's Town Business Association; Greg Culhane and the Evergreen residents at Lake Michelle; Pippa Hudson, Amy-Rae Rispel and the CapeTalk team; Craig Irving and the Hiddingh Community Network; Gail Jennings and the Noordhoek Environmental Action Group; Jodi Allemeier, Diane Awerbuck, Mark Backhouse, Dianne Bayley, Erina Botha, Ted Botha, Ryan Brown of *The Christian Science Monitor*, Edgar Bullen, Phillipa Chitehwe, Mary-Anne Cunningham, Sally Dicey, Jane Dicey, Nerine Dorman, Karima El Abodi and Nafees Mahmud of *TRT World News*, Craig and Bridget Farham, Louise Ferreira, Lesley Green, Jane Griffiths,

Georgina Guedes, Dennis Human, Anthony and Stanley Hungwe, Liesl Jobson of Fetola/Groundswell Africa, Leonie Joubert, Megan Kerr, Jade Khoury of Low Impact Living, Rupert Koopman, Candice Kotze, Rosa Krauss, Helen Laurenson, David Le Page of Fossil Free South Africa, Jamila Maulidi, Mary-Anne Mngomezulu, Pam Morton, Amy Moses, Sue Mottram, Fiona Nicolson, Penny Owens, Kate Noir, Robin Palmer, Justin Phillips, Christine Præsttun of the Norwegian Broadcasting Corporation, Glen Retief, Johan Ripås of Swedish National Television, Peterson Toscano of Citizens Climate Lobby, Lara-Anne Searle-Barnett, Martin Seemungal of *PBS Newshour*, Mandi Smallhorne Kraft, Sydelle Willow Smith, Chip Snaddon, Sally Swartz, Fiona Tipping, Margie Tromp, Marelise van der Merwe, Bernelle Verster of Future Water, Melissa Volker, Barry and Kai Washansky, Patsy White, Anita Wolfaardt, Merina Wolmarans, Kathy and Lauren Wootton, and the unforgettable Teabag Elf.

I have not been able to find the original and anonymous author of the brilliant list of 30 water-saving tips that started circulating on social media soon after the crisis was officially recognised – you were an inspiration.

This book was Louise Grantham's idea, and I've loved working with her appropriately named Bookstorm team: smart, swift and super-professional. Thanks to Nicola van Rooyen for repping this book when it was still a gleam in Louise's eye, to Russell Clarke for painless editing and production, and an undentable sense of humour, to Wesley Thompson for bright-eyed proofreading, and to Marius Roux for turning text into such beautifully crafted pages. The design is all his, and I love it.

Paige Nick staggered off two long-haul flights after all the exhaustion of transcontinental relocation, sat right down and knocked my manuscript, then a messy ramble, into shape in 24 hours. Thank you for your insistence that "your professional be professional". I fall upon your neck in gratitude.

Rodney and Dinah Moffett, my parents and life-long green role models, read the final draft of the manuscript and made helpful suggestions, not all of which there was space and time to implement.

(Composting is indeed a tad more complex than I make it sound.) All errors are mine alone.

Finally, I wrote almost every word of this book with a large three-legged cat rolling around on my keyboard, demanding love. So thank you, Boychik, for hindering me every step of the way.

NOTES